BIOLOGY PRACTICAL GUIDE 7

ECOLOGY

Revised Nuffield Advanced Science
Published for the Nuffield–Chelsea Curriculum Trust
by Longman Group Limited

CONTENTS

Introduction   page 1

Chapter 26   The organism and its environment   5
26A   A qualitative study of a community   5
26B   The effects of different water regimes on plant growth   6
26C   The relationship between nettle distribution and soil phosphate   9
26D   The autecology of woodlice   11

Chapter 27   Organisms and their biotic environments   18
27A   Intra-specific competition in *Drosophila*   18
27B   Inter-specific competition between two *Lemna* species   22
27C   Inter-specific interaction between clover and grass   24
27D   The holly leaf miner (*Phytomyza ilicis*) and its parasitic insects   27
27E   A study of inter-specific associations   31
27F   A grassland survey using sampling techniques   34

Chapter 28   Population dynamics   37
28A   Sampling methods for small invertebrates   37
28B   Capture–mark–recapture technique   39
28C   Population growth in *Chlorella*   41
28D   Population regulation in the water flea, *Daphnia*   46
28E   The responses of a predator to changes in the number of its prey   49

Chapter 29   Communities and ecosystems   51
29A   A quantitative study of an ecosystem   51
29B   The energetics of the stick insect (*Carausius morosus*)   57
29C   A study of decomposer organisms in the soil   59
29D   A comparison of the growth of tolerant and non-tolerant seedlings when exposed to metal ions   61

INTRODUCTION

The practical investigations in this *Guide* relate to the topics covered in *Study guide II*, Part Four, 'Ecology and evolution', Chapters 26 to 29. Cross references to the *Study guide* are given.

(Note that there are no practical investigations accompanying Chapter 30, 'Evolution'.)

Chapter 26   **THE ORGANISM AND ITS ENVIRONMENT**

Investigation 26A   A qualitative study of a community. (*Study guide* 26.1 'Introduction to ecology' and 26.2 'Making and testing hypotheses'.)
The constituent organisms of the community are observed and identified, and distribution patterns of two species are examined.

Investigation 26B   The effects of different water regimes on plant growth. (*Study guide* 26.3 'Factors which influence the distribution patterns of organisms' and 26.5 'Soil factors'.)
The effects of waterlogging on grassland species are investigated.

Investigation 26C   The relationship between nettle distribution and soil phosphate. (*Study guide* 26.3 'Factors which influence the distribution patterns of organisms' and 26.5 'Soil factors'.)

Investigation 26D   The autecology of woodlice. (*Study guide* 26.3 'Factors which influence the distribution patterns of organisms' and 26.6 'Air factors'.)
The preferences of woodlice for different shelter sites are studied. Choice chambers are used to examine aspects of woodlouse behaviour that tend to minimize water loss.

Chapter 27   **ORGANISMS AND THEIR BIOTIC ENVIRONMENTS**

Investigation 27A   Intra-specific competition in *Drosophila*. (*Study guide* 27.1 'Competition'; Study item 27.12 'Shepherd's purse (experiment design)'.)
Competition and selection are observed between vestigial-winged and normal-winged flies.
Investigation 27B   Inter-specific competition between two *Lemna* species. (*Study guide* 27.1 'Competition'; Study item 27.13 'Competition between two flour beetle species in a limited environment'.)

The growth patterns of two species of duckweed are compared, in both single species and mixed cultures.

Investigation 27C Inter-specific interaction between clover and grass. (*Study guide* 27.1 'Competition'; Study item 27.13 'Competition between two flour beetles in a limited environment'.)
The dynamics of the colonization of grassland by clover are described, and the effects of removing one species and of nitrogenous fertilizer are investigated.

Investigation 27D The holly leaf miner (*Phytomyza ilicis*) and its parasitic insects. (*Study guide* 27.2 'Interactions apart from competition' and 27.4 'Parasitism'.)
The common parasites of *Phytomyza* are identified, and their effects on the population are assessed.

Investigation 27E A study of inter-specific associations. (*Study guide* 27.2 'Interactions apart from competition', 27.4 'Parasitism', 27.5 'Commensalism', and 27.6 'Mutualism'.)
Parasitic, commensal, and mutualistic associations of organisms are observed and classified.

Investigation 27F A grassland survey using sampling techniques. (*Study guide* 27.7 'Dispersal and other historical factors'; Study item 27.71 'Dispersal onto Surtsey'.)
Random sampling of grassland communities with a quadrat frame is used to study distribution patterns of species. The differences between communities are related to environmental factors.

Chapter 28 POPULATION DYNAMICS

Investigation 28A Sampling methods for small invertebrates. (*Study guide* 28.1 'How can population size be assessed?')
A range of methods is used to trap the invertebrates of two different habitats. The animals are identified and the populations are compared.

Investigation 28B Capture–mark–recapture technique. (*Study guide* 28.1 'How can population size be assessed?'; Study item 28.11 'The capture–mark–recapture method'.)
The technique is used to estimate the size of a population of small mobile animals.

Investigation 28C Population growth in *Chlorella*. (*Study guide* 28.2 'Population growth'; Study item 28.21 'The growth of a population of yeast'. *Study guide* 28.3 'Population limitation'; Study items 28.31 'The population dynamics of the great tit (*Parus major*) in Marley Wood, Wytham Woods, near Oxford' and 28.32 'Density-dependent, density-independent, and inverse density-dependent factors'.)

Colorimetry is used to monitor the phases of growth of an algal population.

Investigation 28D Population regulation in the water flea, *Daphnia*. (*Study guide* 28.3 'Population limitation'; Study item 28.32 'Density-dependent, density-independent, and inverse density-dependent factors'.)
The effects of density-independent and density-dependent factors on *Daphnia* populations are investigated and compared.

Investigation 28E The responses of a predator to changes in the number of its prey. (*Study guide* 28.3 'Population limitation; Study item 28.32 'Density-dependent, density-independent, and inverse density-dependent factors'.)
A simulation exercise models the effects on a prey population size as a single predator seeks a variable density of prey.

Chapter 29 **COMMUNITIES AND ECOSYSTEMS**

Investigation 29A A quantitative study of an ecosystem. (*Study guide*. Study item 29.11 'Trophic levels and food webs'; Study item 29.12 'Ecological pyramids'.)
The organisms in a model aquatic system are identified and observed, and a food web and a pyramid of biomass are constructed for the system.

Investigation 29B The energetics of the stick insect (*Carausius morosus*). (*Study guide* 29.1 'Introduction'; Study item 29.12 'Ecological pyramids'; *Study guide* 29.3 'Energy flow'; Study item 29.33 'The efficiency of energy transfer between trophic levels'.)
The energy in the food eaten by a stick insect during one week is measured, and the fates of the energy determined.

Investigation 29C A study of decomposer organisms in the soil. (*Study guide* 29.3 'Energy flow' and 29.31 'Decay and decomposition'.)
The processes of decay, including the organisms and the food chains or food web involved, are studied by observing colonies in artificial media in micro-tubes.

Investigation 29D A comparison of the growth of tolerant and non-tolerant seedlings when exposed to metal ions. (*Study guide* 29.5 'Pollution'.)

SAFETY

In these *Practical guides*, we have used the internationally accepted signs given below to show when you should pay special attention to safety.

 highly flammable

 take care! (general warning)

 explosive

 risk of electric shock

 toxic

 naked flames prohibited

 corrosive

 wear eye protection

 radioactive

 wear hand protection

A note for users of this *Practical guide*

The instructions given for the investigations are intended for use as guidelines only. We hope that you will modify and extend the techniques that have been described to meet your own requirements. Other organisms should certainly be tried, depending on what is most readily available. Many of these investigations lend themselves to further work in a Project.

It may not always be possible, for various reasons, for you to do a particular practical investigation. A study of data from another source is perfectly acceptable in such a case.

Many ecological investigations require, and indeed should be planned to allow, statistical analysis of the results. The methods of significance testing mentioned in this *Practical guide* are described in a separate booklet in the Revised Nuffield Advanced Biology series, *Mathematics for biologists*.

CHAPTER 26

THE ORGANISM AND ITS ENVIRONMENT

INVESTIGATION
26A A qualitative study of a community

(*Study guide* 26.1 'Introduction to ecology' and 26.2 'Making and testing hypotheses'.)

A clearly defined area, such as a small pond or an aquarium, supports a community – that is, an interacting group – of organisms. The community usually contains plants, animals, fungi, and prokaryotes. The study of communities usually begins with the description and identification of the organisms.

Procedure

1   Select a small pond, part of a stream, or an artificially constructed ecosystem such as an aquarium.

2   Identify the main types of vegetation within the habitat. For example, in an aquatic habitat such as this there may be distinct patches of floating plants, rooted submerged plants, and rooted plants emerging from the surface of the water. Sketch a plan of the area to show the distribution pattern of these main vegetation zones. If there is an obvious vertical zonation, sketch a vertical section through the water as well, showing the main plants at each level.

3   Collect specimens of as many different organisms as you can and identify them (see the list of aids to the identification of organisms at the end of this book). To obtain as wide a range of organisms as possible, work through each of the vegetation types you defined, as well as the water at the top and the mud at the bottom of your watery habitat. If you are dealing with a natural community, especially on a nature reserve or a Site of Special Scientific Interest, be careful to collect only limited quantities of any plant species that your teacher points out to you.

4   Make a list of the species you find and, if any of them seems to be confined to a particular micro-habitat, mark its position on your plan.

5   Now concentrate your attention on two species only. Examine their distribution patterns. Mark these patterns on the plan and cross-section you have drawn.

6   Record the abundance of the two species in each part of the habitat by using a scale of abundance such as A = abundant, C = common, F = frequent, O = occasional, R = rare. You could use a numerical scale, with points 1 to 10, if you are able to make finer distinctions.

7 Use reference sources, such as those listed at the end of the book, to find out about the life cycles, diet, mode of feeding, and usual habitat of the two species whose distribution pattern you investigated.

8 Construct your own simplified key to one group of organisms you have found, using external features only. Do not forget that size, shape, and colour may show great variation within a species, forming part of the spectrum of continuous variation. Such features are usually of limited value in the construction of keys.

Questions

a *Did you find that the two species you selected each had a characteristic distribution pattern, occurring in some places, and being absent from others? How did these observations link up with your knowledge of the mode(s) of life of these organisms?*

b *Describe any problems you encountered in the identification of the animals and plants which you found.*

c *What characteristics, besides structural external features showing discontinuous variation, can be used in the construction of keys?*

d *What use could be made of characteristics showing continuous variation in the construction of keys?*

INVESTIGATION
26B **The effects of different water regimes on plant growth**

(*Study guide* 26.3 'Factors which influence the distribution patterns of organisms' and 26.5 'Soil factors'.)

Plant species are sensitive to changes in environment. Some species have very specific soil requirements (*Study guide II*, Section 26.5 and table (S)58). A species which is an effective competitor in one environment may be ineffective when its environment is changed.

One important factor which affects the species composition of plant communities is the amount of water in the soil. A high soil water content is often indicated by the presence of sedges and rushes in the community. Many water plants (for example, sweet flag, *Acorus calamus*) can be grown on dry soil in gardens, however, provided that more effective competitors are excluded.

The purpose of this experiment is to assess the effects of waterlogging on various grassland species which are not normally abundant under wet conditions. Can the composition of the community, or at least the growth rates of the species, be changed by altering this 'single' environmental factor?

In this investigation the water content of the soil under grass turves is altered by changing the water level, or water table, surrounding them.

Procedure

1 Cut three turves 20 × 20 cm and 10–15 cm thick from a piece of close-cut grassland which contains both grasses (monocotyledonous) and broad-leaved (dicotyledonous) species. Try to select adjacent turves from the same site, as similar in species composition as possible.
2 Record the plant species present in each turf by means of sketches. Count and record the number of individuals of each species in each turf.
3 Select two common species, say one dicotyledonous plant and one monocotyledonous plant. Map the positions in the tray of ten individuals of each of these species. Measure in mm, and record, the length of the longest leaf on each of these twenty plants. After three weeks the longest leaf lengths of the same plants will be measured again.
4 Place each turf in a separate seed tray, preferably 15 cm deep with no drainage holes. In one corner of each tray set up a siphon tube, positioning the tubes as shown in *figure 1.*
5 Water the trays well and siphon the excess water away by using the three-way tap and syringe. Fill the siphon tube with the tap in the

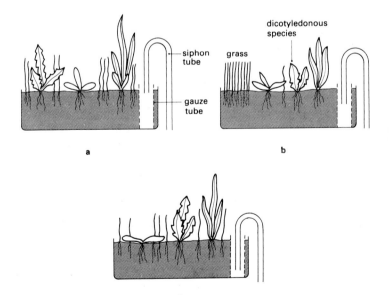

Figure 1
Vertical section through the experiment, showing the seed trays with the turves inserted.

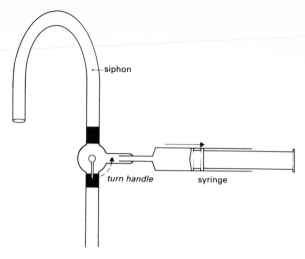

Figure 2
The method for starting a siphon, using a syringe. First draw the water down to the tap by pulling out the syringe barrel. Then turn the handle of the three-way tap upwards and to the right, to start the siphon.

position shown in *figure 2*. Then open the tube and draw off the excess water. Keep the trays in a well-lit place in the laboratory, or out of doors, and as cool as possible during the experiment.

6   At least once a day the water levels should be checked and the turves watered as explained above. This should ensure that no significant change in the water levels occurs during the three-week period of the experiment.

7   Leave for three weeks, and then measure again the longest leaves on each of the twenty plants previously selected in each tray (see **3** above).

8   Examine your previous notes and, for each turf, record   *a* which species have spread and   *b* how many individuals of each species have died.

9   Tabulate the results to show the effect of the three different treatments on the growth of the grasses and dicotyledonous plants. Your table should have three vertical columns, one for each treatment, and a row for each set of measurements you have made.

10 Test your results statistically. First, have the leaf lengths increased significantly in each treatment? Use a Mann–Whitney U-test (see *Mathematics for biologists*) to test the ten leaf lengths in a particular treatment at the start of the experiment against the leaf lengths at the end. If you are sure that the same plant individuals have been sampled at the beginning and end of the experiment, you are justified in carrying out the more precise Wilcoxon signed-ranks test for matched

pairs. What do your results indicate? Second, use a Mann–Whitney U-test to discover whether or not the increases in leaf length are significantly different between treatments.

Questions

a   *What effects seen during the experiment do you think are caused by the different water tables in the trays? Give reasons for your answers.*

b   *What other soil factors may have changed compared with the turves in their natural environment?*

c   *What are the problems in using leaf length as a measure of plant performance?*

d   *What would be a better measure of performance of the plants than any of the other features you have measured or observed?*

e   *Suggest three other observations or measurements which would be useful in estimating the performance of plants.*

f   *What practical importance could your observations have for the gardener?*

g   *Describe a suitable method for investigating the effect of one other soil factor, such as soil nutrient level, on a grassland community.*

INVESTIGATION
26C   The relationship between nettle distribution and soil phosphate

(*Study guide* 26.3 'Factors which influence the distribution patterns of organisms' and 26.5 'Soil factors'.)

The distribution pattern of a species is controlled by various factors. Some of these factors may be identified. The first step is to find a relationship in the field between the distribution pattern or abundance of the species and the level of a measurable environmental factor or factors. This does not prove, of course, that the factor – say the nitrate concentration in the soil – affects the pattern of the species, but the clue is worth following up by field or laboratory experiments. These experiments may establish beyond reasonable doubt a causal connection between the factor and the distribution pattern.

This investigation is to test in the field the hypothesis that nettles might be influenced by levels of soil phosphate. If you find a positive or negative association it may be worth trying to establish a causal connection by field or laboratory experiments, either as a class practical, or in project work.

Procedure

1   Select an area of at least 20×20 m which has nettles growing on part of it.

2   Map the position of the main plant communities in the area by pacing it out and making a sketch map.

3   Collect and label four separate soil samples from each of the main communities. Each sample should be sufficient to fill a small specimen tube and should consist of a mixture of soil from the surface down to a depth of 10 cm.

Instructions **4–8** explain how to prepare a standard curve for the estimation of phosphate concentration in the soil samples. This can be done simultaneously with the soil analysis (instructions **9–13**).

4   Label six 35 cm³ volumetric flasks 2, 5, 10, 15, 20, and 25 µgP. Add to them respectively 1, 2.5, 5, 7.5, 10, and 12.5 cm³ of the dilute phosphate solution (potassium dihydrogen phosphate) provided, which contains 2 µg of phosphorus per cm³.

5   Add 5 cm³ of sodium hydrogen carbonate solution to each flask.

6   Add 5 cm³ of ammonium para-molybdate solution to each flask. There will be a rapid evolution of carbon dioxide. Afterwards, shake the mixture and make it up to about 22 cm³ with distilled water.

7   Add 1 cm³ of dilute tin(II) chloride (stannous chloride) solution to each flask, and make up to the mark with distilled water. Put the top on the flasks and mix the contents immediately. Leave the flasks to stand for exactly ten minutes. Pour some liquid from the 25 µgP flask into a sample tube. Place the sample tube into a colorimeter fitted with a red filter (see *Practical guide 2*, investigation 5C). Adjust the meter reading to give a full-scale deflection. Pour each of the other samples in turn into the same clean sample tube, and record their colorimeter readings, as quickly as possible.

8   Make a table of results and plot them on log-normal graph paper. Plot the percentage transmittance on the linear scale and the concentration of phosphate on the logarithmic scale.

Instructions **9–13** describe how to analyse each soil sample to find the amount of phosphate it contains.

9   Place 5 g of moist soil with 100 cm³ of the sodium hydrogen carbonate solution provided into a 250 cm³ conical flask. Cork the flask and shake for about fifteen minutes, either by hand or in a mechanical shaker. The shaking time does not matter much provided that it is the same for all the soil samples.

10 Filter the soil mixture through a filter paper and collect the clear filtrate.

11 Withdraw 5 cm³ of the filtrate and place it in a 25 cm³ volumetric flask.

Then, using a 5 cm³ syringe, slowly add 5 cm³ of stock ammonium para-molybdate solution. After the evolution of carbon dioxide, shake the mixture and make it up to about 22 cm³ with distilled water.

12 Add 1 cm³ of dilute tin(II) chloride (stannous chloride) solution, and make up to the mark with distilled water. Put the top on the flask and mix the contents immediately. Leave it to stand for exactly 10 minutes.

13 Measure the transmittance using a colorimeter with a red filter (ideally 660 nm) and record the results.

14 Read off the concentration of phosphate in the sample from the standard curve prepared earlier.

15 Average the phosphate concentration in the four samples taken from beneath each vegetation type. If you feel that there is a real difference in phosphate concentrations between soils from beneath any two vegetation types, use a Student's *t*-test to test the significance of the difference.

Questions

a *Describe any visible factors which might affect nettle distribution in the area you studied.*

b *Are the phosphate concentrations in nettle-infested areas significantly different from those in the other areas?*

c *Why are phosphates important for the growth of plants such as nettles?*

d *What other nutrient ions may also differ in concentration in the areas with and without nettles?*

INVESTIGATION
26D The autecology of woodlice

(*Study guide* 26.3 'Factors which influence the distribution patterns of organisms' and 26.6 'Air factors'.)

Woodlice are the only fully terrestrial representatives of the marine class of arthropods, the Crustacea. They all occupy habitats where water loss is minimized, in crevices, between stones, beneath bark or seaweed, and in other dark and damp places. Each species occupies a distinct habitat, however. In this investigation field observations and laboratory practicals are used to investigate some aspects of woodlouse ecology.

Four common species are illustrated (*figure 3*) as they have a fairly wide distribution. Other species may be locally abundant, however.

Oniscus asellus is the most abundant species. It is 15 mm long, with a

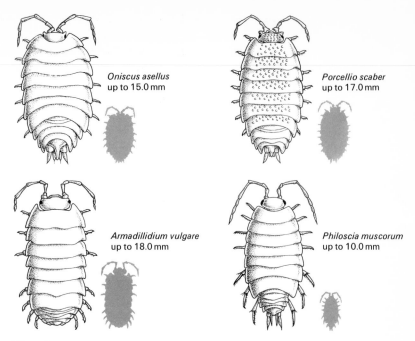

Figure 3
Common woodlice. The silhouettes show *relative*, not actual, shapes and sizes.

distinctive grey glossy body. It has lateral lobes to the head and a telson which ends in a long point. The end of the antenna, called the flagellum, has three sections.

Porcellio scaber is distinguished from *Oniscus* by the flagellum of two segments. It is more flattened dorso-ventrally than *Oniscus* and has a tuberculate (lumpy) appearance. The colour is usually a uniform dark slate grey, while the head has the same kind of lateral projections as that of *Oniscus*.

Armadillidium vulgare is similar to *Porcellio* in having a dark coloration, but it is usually shiny and lacks the projecting exopodites of the uropods found in *Porcellio*. It is strongly curved in section and can roll into a ball. This species is found only where there is chalk or on the sea coast, and is known as the common pill-bug.

Philoscia muscorum is a species with a mainly southern distribution. It is brown in colour, with a darker head. Lighter median strips are characteristic. It is smaller than the other species described and is able to tolerate drier conditions, being found in dune grassland as well as in other habitats.

The biology of the common species is primarily concerned with the problems of water conservation. When the temperature is low and the relative humidity increases the activity becomes greater. Changes in the humidity levels can be shown to have seasonal and other circadian rhythms, and in some species photo-negative reactions are also related to the high rate at which these organisms lose water.

The aim of this investigation is to explain the distribution patterns of woodlice in the field in terms of their behaviour in laboratory experiments.

Procedure
Locate several colonies of woodlice. They may be either naturally occurring colonies or colonies kept in covered troughs in the laboratory. Try to have at least two species present.

1 Estimate the population size using the capture–mark–recapture method (investigation 28B).

2 Try to determine the extent to which woodlice return to the same shelter:

a Collect as many woodlice as you can from two adjacent shelter sites (colonies) not more than a metre apart, using a spoon, a large pooter, or a paintbrush.

b Keep the individuals from each colony in separate screw-top containers.

c Mark the individuals with a coded mark for each colony. A felt-tipped permanent spirit marker is suitable for this purpose.

d Release the animals at a point about equidistant from the two colonies.

e Record the numbers of each of the two kinds of marked individuals in each of the two shelter sites after twenty-four hours. Keep recording in a similar manner for two or three days.

Questions

a *Do the data suggest that woodlice have a preferred shelter site?*

b *If the number of marked woodlice recovered in such a field experiment declined rapidly, what might you suspect?*

c *If there were a large number of unsuspected shelter sites in the area under investigation, and the woodlice showed no preference for the home site, what pattern of results might you expect?*

d *How would you change the experiment so that interference between individuals from different home sites was avoided?*

e *What data, other than those referred to in the instructions for the experiment, could be used to interpret the results of the experiment, and how would you obtain and record such data?*

To understand the distribution of the woodlice, it is necessary to keep them in the laboratory and to devise suitable experiments to investigate the effects of one environmental factor.

Water loss in clustered and isolated individuals

a

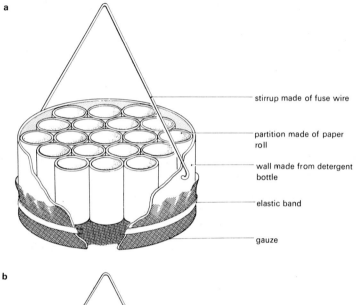

stirrup made of fuse wire

partition made of paper roll

wall made from detergent bottle

elastic band

gauze

b

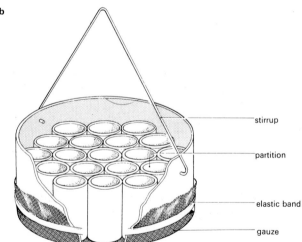

stirrup

partition

elastic band

gauze

Figure 4
Apparatus for measuring water loss in clustered and in isolated woodlice.
a Woodlice allowed to cluster.
b Woodlice isolated.

Procedure

1 You are provided with two containers for holding woodlice. One allows the woodlice to cluster, and the other has partitions which keep the woodlice apart.

2 Select fifty individuals of approximately the same size, and place twenty-five in each container.

3 Place both containers into a larger container lined with moist paper, and keep them at a high level of humidity for ten minutes.

4 Weigh each container to the nearest 0.001 g and record the mass.

5 Suspend the containers in the air in a dry laboratory for 15 minutes (*figure 4*).

6 Reweigh, as before, and record the mass.

7 Calculate the loss of mass in each of the containers and compare them.

Questions

a ***Do your results indicate that the clustered woodlice lose mass more slowly than the isolated ones?***

b ***Suggest a reason for the results which you have obtained.***

c ***Do woodlice cluster together in their natural habitat?***

d ***What bearing do your results have on your field observations?***

Isolating aspects of behaviour which lead to water loss

Procedure

1 Set up choice chambers (*figure 5*) containing thoroughly wetted paper towel lining the bottom of the chamber on one side, and a suitable drying agent such as calcium chloride on the other. Place a piece of cobalt thiocyanate paper as shown in the diagram, to indicate the presence of a humidity gradient.

2 Leave for ten minutes to establish a relatively stable humidity gradient.

3 Insert five woodlice on the dry side and five on the wet side of the chamber. Seal up the holes with sellotape and cover up the choice chamber to exclude light.

4 Leave the apparatus for about five minutes.

5 Remove the cover, and count and record the number of woodlice on each of the two areas.

6 Replace the woodlice with new specimens and repeat the stages **1–5**, rotating the chamber through 180°. If the number of woodlice available is limited, the original sample can be used again after about an hour of recovery in the original container. Light penetration into

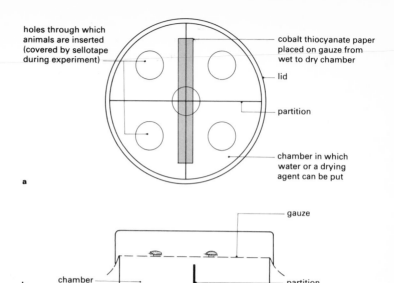

holes through which
animals are inserted
(covered by sellotape
during experiment)

cobalt thiocyanate paper
placed on gauze from
wet to dry chamber

lid

partition

chamber in which
water or a drying
agent can be put

a

gauze

chamber

partition

b

Figure 5
A choice chamber. The fine gauze, on which the animals walk, is stretched horizontally across the container. The space beneath it is divided by partitions into four compartments, into which water or drying agents can be put to create humidity gradients.
a View from above.
b Side view.

the chamber, temperature variations, and surface irregularities are eliminated as factors influencing distribution, by rotating the chamber at the end of each determination.

7　Repeat the procedure as often as possible.

Questions

a　*What preference, if any, did the woodlice have in the humidity gradient?*

b　*Did all the woodlice show a similar response to the same conditions? Can you explain any differences which occurred?*

c　*Why is a standard pre-treatment laid down for the woodlice?*

d　*Describe how you would determine the effect of pre-treatment on the behaviour of woodlice.*

e　*Under what circumstances would you expect differences in behaviour to occur? Bear in mind how the selection of the experimental animals is carried out.*

To explain the mechanisms by which woodlice locate and remain in humid atmospheres, further experiments have to be carried out. There are two major response mechanisms: kineses and taxes (see *Study guide I*, section 12.2 and *Practical guide 4*, investigations 12A and 12B). A *kinesis* is a response to a stimulus involving a change in the level of activity. A *taxis* involves only a change in orientation in relation to the stimulus. The woodlice carry out random turning movements in the absence of known stimuli. Such movements are speeded up by the presence of water and this reaction is an example of a kinesis.

Suggestions for further reading

BENNETT, D. P. and HUMPHREYS, D. A. *Introduction to field biology*. 3rd edn. Arnold, 1980.

CAMPBELL, R. C. *Statistical methods in biology*. 2nd edn. Cambridge University Press, 1974.

COLLINS, M. *Urban ecology*. Cambridge University Press, 1984.

DARLINGTON, A. and LEADLEY -BROWN, A. *One approach to ecology*. Longman, 1975.

DOWDESWELL, W. H. *Ecology: principles and practice*. Heinemann Educational, 1984.

ETHERINGTON, J. R. Studies in Biology No. 141, *Wetland ecology*. Arnold, 1983.

Revised Nuffield Advanced Science: Biology, *Mathematics for Biologists*. Longman, 1987.

The following are useful aids to the identification of organisms:

CLOUDSLEY-THOMPSON, J. L. and SANKEY, J. *Land invertebrates*. Methuen, 1961.

DAVIS, B. N. K. *Insects on nettles*. Cambridge University Press, 1983.

SUTTON, S. *Woodlice*. Ginn, 1972.

CHAPTER 27

ORGANISMS AND THEIR BIOTIC ENVIRONMENTS

INVESTIGATION
27A **Intra-specific competition in *Drosophila***

(*Study guide* 27.1 'Competition'; Study item 27.12 'Shepherd's purse (experiment design)'.)

Interactions can occur within populations of a single species as well as between populations of different species. If the interaction is one resulting from a shortage of some resource (for example, a nutrient, light, water, breeding sites), then competition is said to occur.

Intra-specific competition, that is, competition within a population of the same species, is likely to be especially severe since all the individuals require the same set of resources. Most populations exhibit genetic variation. Competition may increase the chances of survival of some genotypes over others, that is, competition can result in selection. To study this effect, rapidly breeding populations showing distinct phenotypes are required. The fruit fly *Drosophila* is ideal because of its rapid life cycle and the ease with which it can be reared.

The gene vestigial wing (vg) produces small distorted wings in adult flies. It is recessive to the wild-type gene (vg$^+$) which produces normal wings. Competition and selection can be observed between the vestigial-winged and the normal-winged phenotypes.

Procedure
The following procedure will enable you to produce three specimen tubes, containing respectively:
A ten pairs of wild-type flies (vg$^+$ vg$^+$),
B five pairs of wild-type flies (vg$^+$ vg$^+$) and five pairs of vestigial-winged flies (vg vg).
C ten pairs of vestigial-winged flies (vg vg).

1 Assemble the etherizer and have all the components of the glass jar population cages (*figure 6*) at hand. You will not have time to look for apparatus once you have anaesthetized the flies.

2 Wet the cottonwool in the etherizer (*figure 7*) with a little ether. It must not drip into the tube or the flies will be killed.

3 Tap the vestigial-winged fly culture sharply down on to the bench or other firm surface to dislodge flies from the screw-cap or plug. Pull out the plug and knock the flies into the etherizer funnel. Do not try to harvest all the flies from the bottle at once. Replace the culture bottle plug quickly.

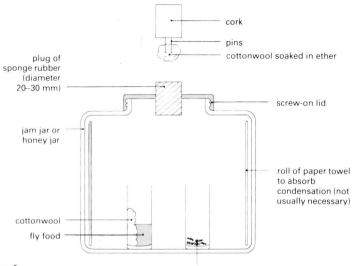

Figure 6
A glass jar population cage for *Drosophila*.

cork

pins

cottonwool soaked in ether

plug of sponge rubber (diameter 20–30 mm)

screw-on lid

jam jar or honey jar

roll of paper towel to absorb condensation (not usually necessary)

cottonwool

fly food

unconscious flies

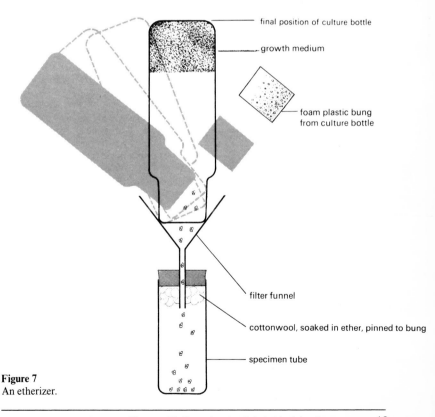

Figure 7
An etherizer.

final position of culture bottle

growth medium

foam plastic bung from culture bottle

filter funnel

cottonwool, soaked in ether, pinned to bung

specimen tube

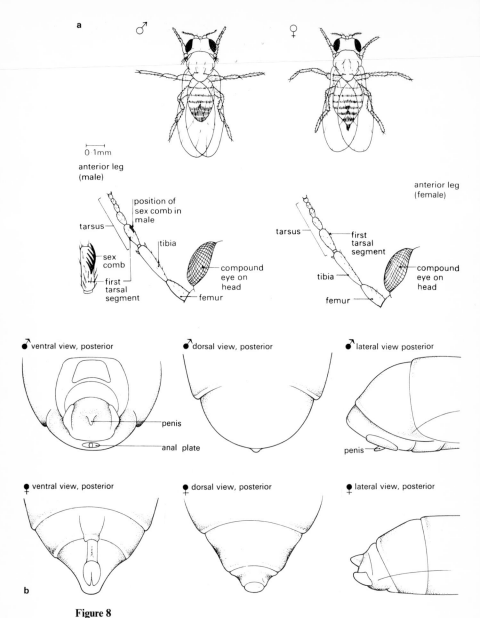

Figure 8

Male and female *Drosophila*.

a Whole flies, low power, showing anterior legs and sex comb of male.

b Tips of abdomens, higher power, showing ventral, dorsal, and lateral views.

4   Wait until all the flies in the etherizer are motionless. Tap the tube to see whether or not they will fall to the bottom.

5   Pour the flies on to the white tile.

6   Move the flies about gently with a paint brush. Observe with the dissecting lens or microscope and sort into males and females using *figure 8* as a guide. If any flies become active, tip or sweep them into the etherizer funnel for a second short dose of ether.

7   Place ten pairs of vestigial-winged flies into the empty specimen tube C, and five pairs into the tube B.

8   Repeat instructions **3** to **6** with the wild-type culture.

9   Place ten pairs of wild-type flies in tube A and five pairs in tube B.

10   Write the date on three tubes of fly food. Add an equal mass of dried yeast to each tube (three or four granules will do).

11   Place one of these dated food tubes into each of the glass jar population cages. Label the cages A PURE WILD, B MIXED, and C PURE VESTIGIAL.

12   Place the correct tube of flies in each cage and close the cage, making sure that the rubber bung fits.

13   Incubate the three cages at 25 °C and check them twice weekly.

14   At agreed time intervals, which will depend on the exact temperature of the incubator and other factors, anaesthetize the flies, score them as wild-type or vestigial-winged, and replenish the food supply. Treat each of the three glass jars as follows:

a Carefully remove the rubber bung and prevent the flies from escaping by sliding a piece of card over the opening.

b Substitute a cork with ether-wetted cottonwool pinned on to it.

c Turn the jar on its side so that the flies will not fall on to sticky medium.

d As soon as the flies are motionless, remove the food tubes with forceps.

e Tip the flies on to a tile and blow the ether vapour off the tubes and from the jar. If ether fumes remain, there is an increased risk of sterility in the flies.

f Count the flies and score them as wild-type or vestigial-winged. Enter the result in a table.

g Dead flies have their wings held perpendicularly to the body and have a shrivelled appearance. Discard them.

h Place all the live, motionless flies in an empty specimen tube.

i If the food has dried in any of the food tubes so that it has contracted away from the glass, add up to 2 cm³ of water to *all* the tubes.

j Prepare a fresh, dated tube of food to which a known mass of yeast has been added. Place it in the cage and check that no ether fumes remain.

k Reassemble the cage, excluding the oldest tube of food if there is no room for it. Your jar should now contain one empty tube of motionless flies and up to four tubes of food of various ages.

15 Construct graphs of the numbers of flies of the two phenotypes in each of the population cages plotted against time. Discontinue the observations when clear trends have emerged.

Questions

a *What was the purpose of setting up the two populations of pure cultures?*

b *The gene for vestigial wings is recessive to the wild-type gene. How may this influence the relative numbers of vestigial-winged and normal-winged individuals in the population in jar B, the mixed population? Do the results support your hypothesis?*

c *Three possible hypotheses to explain the decline in the frequency of the vestigial-winged flies are listed below. Suggest some experiments which would distinguish between these hypotheses.*
   1 Vestigial-winged flies cannot efficiently compete for certain resources with wild-type flies.
   2 Vestigial-winged flies are less able to survive some environmental stresses, such as unsuitable humidity, than the wild-type flies.
   3 Fruit flies use their wings during courtship. A female need only mate once in her life. A male may mate several times. Vestigial-winged males might be at a disadvantage in courtship.

d *It has been reported that vestigial-winged fruit flies are frequent in the wild on very small islands in the Pacific Ocean. Suggest a hypothesis to explain this finding.*

INVESTIGATION
27B Inter-specific competition between two *Lemna* species

(*Study guide* 27.1 'Competition'; Study item 27.13 'Competition between two flour beetle species in a limited environment'.)

When two species occupy the same habitat, they will interact with one another. This interaction may be mutually beneficial, mutually disadvantageous, or favouring only one of the species. In order to study one such interaction two species of duckweed, *Lemna minor* and *Lemna trisulca*, which are often found growing together, can be grown under laboratory conditions where the nature of the interactions can be monitored in a controlled environment.

Procedure

1 Connect three containers, each capable of holding about 400 cm³ of water, by siphons so that the water flows from a large reservoir of aerated pond water at the rate of about 800 cm³ per day (see *figure 9*). The water can be run to waste or returned to the stock tank by means of a pump or an air-lift.

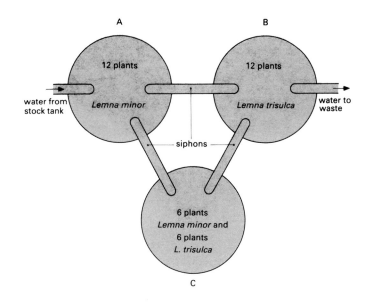

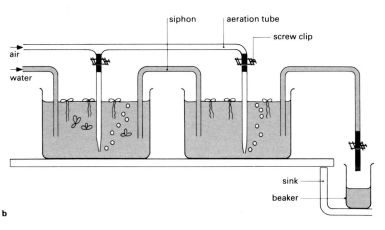

Figure 9
Apparatus for investigating competition between two species of *Lemna*.
a View from above.
b Part of side view.

2 Add the *Lemna* plants in the proportions shown in *figure 9* and set the containers in a sunny place or under fluorescent lights. Aerate at a fairly slow rate through fine tubes. Make sure that the bubbles do not cause the *Lemna* plants to spread up the sides of the tank.

3 Observe the cultures twice a week and count the numbers of 'leaves' (often called 'fronds') of each species in each of the cultures. Transfer the plants carefully using a paint brush so that the fresh containers become the new culture vessels.

Questions

a *Explain briefly how*
1 variables were controlled and
2 the number of living leaves were counted in your investigation.

b *Plot a graph on log-normal graph paper to show the pattern of growth of each of the **Lemna** species when grown in isolation. What evidence did you find of intra-specific interaction during the experiment?*

c *Plot graphs, on linear graph paper, one for each species of **Lemna**, comparing the number of leaves produced in the single species cultures with the number of leaves produced in the mixed culture. In the conditions of this experiment, what evidence do you have that **inter-**specific interaction is a significant factor in controlling the growth of either species of **Lemna**? Explain your answer.*

d *Interaction between plants growing together in a habitat may involve competition for environmental resources. What resources might restrict the growth of **Lemna** in your experiment? Select the resource which you consider to be the main limiting factor and suggest an experimental procedure which you could use to investigate its effect in more detail.*

INVESTIGATION
27C   Inter-specific interaction between clover and grass

(*Study guide* 27.1 'Competition'; Study item 27.13 'Competition between two flour beetles in a limited environment'.)

Many lawns, grass verges, and pastures are colonized by white clover (*Trifolium repens*) which grows amongst the grass. The flowering heads are white and the leaves are each composed of three leaflets. There may be white markings on the leaves, the extent and patterns of which are genotypically determined. The plant is hairless. It spreads by seed and by stems (stolons) which lie on the surface of the ground and produce rootlets at the nodes. Clover is prostrate and dwarfed in well-mown lawns, but amongst rank grass it becomes rampant, with larger leaves and fleshier stolons.

Most lawns contain both grass and clover and, unless the lawn is specially treated, the two species live together for many years. What would happen if one species were removed? Would the remaining partner suffer, or grow to colonize the bare ground? How does it happen that the two can flourish and neither replace the other? These are questions that an ecologist or a gardener who is interested in the dynamics of the community might ask. Two investigations may illuminate the relationship between the clover and the grass on a lawn.

Mechanical removal of one partner

Procedure

1   Map an area of clover and grass 1 m square (*figure 10*). You need not record individual plants unless they are large. Take a photograph of

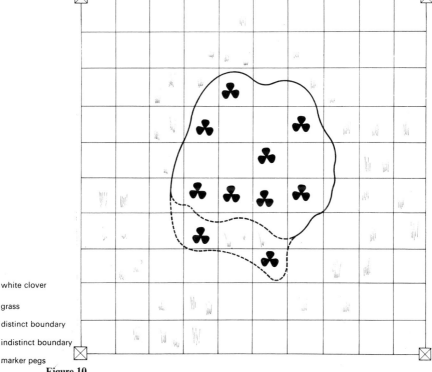

Key

white clover

grass

distinct boundary

indistinct boundary

marker pegs

Figure 10
Typical map of the distribution of clover and grass in a square metre.

the area too. Sink four pegs into the ground, one at each corner, so that you will be able to recognize the marked area – the quadrat – again. Measure and record the distances of two of the four pegs from fixed objects. Note the date.

2 Remove the grass leaves and stems from an area next to a patch of clover within the quadrat. Choose a place where the boundary is well-defined. Remove the clover stolons and clover leaves from another area. Record the cleared areas on your map.

3 Examine the plants from time to time. Remove plants which are neither clover nor grass if they have invaded the cleared area.

4 Keep your map in a safe place when you have finished so that it can be used for comparison in later years.

Questions

a *Does the evidence suggest that in the absence of any one species the other might eventually colonize the lawn?*

b *What sort of change might prevent this from happening?*

c *In what ways does the disturbance change the micro-habitat? How might this affect the interaction between clover and grass?*

Treatment with nitrogenous fertilizer

Very near to the clover, the grass may be greener and more lush, and this suggests that the supply of nitrogen is plentiful. In their root nodules, legumes such as clover harbour bacteria which can fix atmospheric nitrogen, releasing it into the soil as nitrate.

Assess the effect of adding a nitrogen compound, such as ammonium sulphate, to a lawn where clover and grass are growing together.

Procedure

1 Mark out another quadrat 1 m square and map it as before.

2 Dissolve 20 g ammonium sulphate crystals in about 4 dm^3 of water. Apply this solution to the plot with a watering can fitted with a rose.

3 Repeat the treatment after a week, especially if the soil is sandy.

4 Keep records of the treatments alongside the map, and observe any changes in appearance and extent of areas colonized by clover and grass.

Questions

d *Does grass grow more vigorously when extra nitrogen is supplied? If so, what advantage could the increased growth confer on the grass?*

e *Does clover grow better with additional nitrogen? How do you explain the results which you have obtained?*

f *What evidence is there that grass responds in the same way to added nitrogen and to being near clover? If there is a similarity, can you put forward a hypothesis to account for it?*

g *Is there any evidence of competition between clover and grass? If so, for what might they be competing?*

h *Do the results of your experiments justify the conclusion that clover and grass are competing for a limited amount of nitrogen in the soil?*

INVESTIGATION
27D The holly leaf miner *Phytomyza ilicis* and its parasitic insects

(*Study guide* 27.2 'Interactions apart from competition' and 27.4 'Parasitism'.)

Parasitism is a special form of interaction in which a delicate balance usually exists between the number of hosts and the number of parasites. The larvae of the small fly *Phytomyza ilicis* burrow through holly leaves and produce the yellow lines and dark blotches from which the name holly leaf miner is derived. Sometimes more than half the leaves on a tree are attacked. The holly leaf miner is itself a parasite on holly, its host. This fly is parasitized by at least nine different species of wasp (Hymenoptera) which lay their eggs inside the larva or pupa. In this investigation you will find the larvae or pupae of the holly leaf miner, and if they have been parasitized, you will be able to identify the parasite. The life history of the holly leaf miner is shown in table 1.

	Jan	Feb	Mar	Apr	May	Jun	Jly	Aug	Sep	Oct	Nov	Dec
Holly leaf miner (*Phytomyza ilicis*)	L	L	L/P	P	P/A	A/E/L	L	L	L	L	L	L
Chrysocharis gemma	A	A	E/L	L/P	P	P/A	A	A	A	A	A	A
Chrysocharis syma	A	A	A/E/L	E/L	L/P	P/A	P/A	A	A	A	A	A
Sphegigaster flavicornis	A	A	A	A/E/L	L/P	P/A	P/A	A	A	A	A	A

Table 1
The life history of the holly leaf miner and its three commonest parasites.
Key: A = adult, E = egg, L = larva, P = pupa.

The three parasites which occur most frequently are:

1 *Chrysocharis gemma*, which lays its egg into the fly larva. Attacked larvae are pale, dirty yellow compared with the bright, shiny whitish-lemon healthy larvae. The parasite then forms a shiny jet-black pupa which lies free inside the mine. The adult emerges by a small, near round hole, leaving the black pupal skin behind.

2 *Sphegigaster flavicornis* lays its egg on the pupa of the fly. The larva bores into the pupa, feeds, and then pupates. The pupa is black with a bluish tinge, except for antennae, wings, and legs which are light, glassy brown.

3 *Chrysocharis syma* has a similar life cycle to that of *C. gemma*, except that it attacks the fly pupa. Its pupa, inside the fly puparium, is shiny black.

These parasites are illustrated in *figure 11*.

Procedure

1 Collect holly leaves from an infested tree, preferably in May when the total mortality due to parasitism can be assessed. If leaves are collected earlier in the year, a smaller proportion will be parasitized. As far as possible, the leaves should all be of the same age. They can be kept in the refrigerator until required.

2 Estimate the proportion of leaves on the tree attacked by the leaf miner. Select twigs with about twenty leaves. Examine each leaf and record whether or not it is mined; collect the mined leaves and discard the others. Continue examination until about 200 leaves have been collected. Try to select twigs from as wide a range of heights and aspects as possible.

3 In the laboratory open each mine and record the contents in the categories suggested in table 2 with the help of the description in *figure 11* and key in table 3. Mines attacked by birds are torn open and may have roughly triangular beak marks on them.

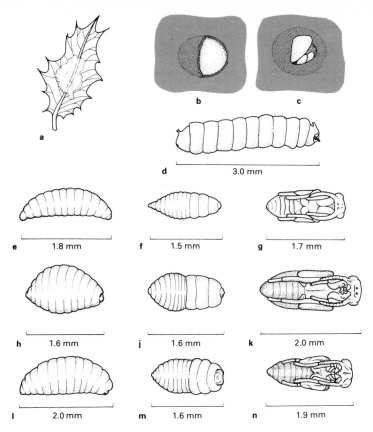

Figure 11

Stages in the life history of the holly leaf miner and its commonest parasites.

a Holly leaf mined by the holly leaf miner (*Phytomyza ilicis*).

b *P. ilicis*. Puparium before emergence of adult from leaf.

c *P. ilicis*. Puparium in leaf after emergence of adult.

d *P. ilicis*. Larva.

e *Chrysocharis gemma*. Mature larva.

f *C. gemma*. Prepupa.

g *C. gemma*. Pupa.

h *Chrysocharis syma*. Mature larva.

j *C. syma*. Prepupa.

k *C. syma*. Pupa.

l *Sphegigaster flavicornis*. Mature larva.

m *S. flavicornis*. Prepupa.

n *S. flavicornis*. Pupa.

From Lewis, T. and Taylor, L. R., Introduction to experimental ecology, *Academic Press, 1967.*

Date of collection	....
Total leaves examined	....
Leaves with mines	....
Leaves without mines	....
Empty pupal cases	....
Dead larvae or pupae	....
Larvae or pupae attacked by *Chrysocharis gemma*	....
Pupae attacked by *C. syma*	....
Pupae attacked by *Sphegigaster*	....
Healthy unparasitized larvae	....
Healthy unparasitized pupae	....

Table 2
A recording sheet for mines made by the holly leaf miner.

1	blotch with hole	**2**
	blotch without hole	**4**
2	hole a large irregular tear	bird predation
	hole triangular, often with flap	*Phytomyza* emerged
	hole circular	**3**
3	hole similar to feeding punctures scattered over leaf	irrelevant?
	neat round hole, no necrotic tissue round it	*Chrysocharis gemma* emerged
	neat round hole, surrounded by dry necrotic oval	*C. symna* or *Sphegigaster* emerged
4	Blotch without hole:	
	large, healthy-looking maggot	*Phytomyza*
	flat, larval skin in brown plant tissue	maggot killed by sucking predator
	black, hymenopterous pupa	*Chrysocharis syma*
	brown puparium	**5**
5	Open the puparium:	
	adult fly within	*Phytomyza*
	empty	*Phytomyza* emerged
	pupa (or cast skin)	**6**
6	pupa or cast skin black	*C. syma*
	blue-black pupa with pale appendages	*Sphegigaster*

Table 3
Identification key for the causes of death in the holly leaf miner. If the blotch has a hole (statement 1, first alternative) the mine should be dissected to confirm that there are no larvae or pupae in the leaf.

Questions

a *What percentage of the holly leaves were affected by the holly leaf miner? Compare your result with those of other groups. How can the differences in the size of infestation be explained?*

b *What effect does the miner have on the holly tree? Were heavily infested trees less vigorous than trees with more healthy leaves?*

c *If the original population of the leaf miner is taken to be the number of leaves with mines, what percentage of them has survived? What percentage has been parasitized?*

d *In one study only 2 per cent of pupae examined in late May contained healthy leaf miners. What impact do the parasites have on the population of the miner? What would happen if the parasites were eliminated?*

INVESTIGATION
27E A study of inter-specific associations

(*Study guide* 27.2 'Interactions apart from competition', 27.4 'Parasitism', 27.5 'Commensalism', and 27.6 'Mutualism'.)

If two species interact so that the life history of one or both of them cannot be completed on its own, then a permanent association of some kind is formed. Such associations can be divided into three major categories.

1 *Commensalism* is an association in which one partner benefits from an association but the other neither benefits nor is harmed.

2 Both organisms benefit in a *mutualistic* association.

3 The most extreme form of association is *parasitism*, where one member of the partnership, the host, is harmed and the parasite forms a permanent association from which it derives positive benefit. The parasite is often modified and specialized for its mode of life and restricted in its ability to lead an independent existence.

If one of the partners in any of these associations is able to live an independent existence it is said to be *facultative*, while the dependent partner is said to be *obligate*. In some cases the association may be so close that the partners form a single structure which is different from either of the partners.

The closest associations of parasitism benefit the parasitic partner because of the proximity of food, shelter, and a place to breed. Factors which limit the success of a parasite include the difficulty of finding a new host and the increased specialization that results from dependence on a single host species.

Procedure
Examine several different associations and try to see if they fit into the classification given in the introduction. After you have studied each example, try to explain the ways in which the organisms interact. In some

cases you may need to consult some of the reference books mentioned in the list at the end of this chapter.

A   Aphid and green plant
1   Observe aphids (greenfly) on a leaf, using a binocular microscope. Describe how they seem to be feeding, and how they are distributed on the surface.
2   Examine leaves for signs of damage, or the formation of galls.
3   Examine the head of an aphid, mounted on a microscope slide.

Questions

a   **What type of association exists between the plant and the aphid?**

b   **Explain your reasons for your answer to part a.**

B   Mosquito and mammal
1   Examine a slide of the head of a mosquito, under the low power of the microscope. Note the way in which the jaws are modified for piercing and sucking.

Questions

a   **What evidence do you see to suggest that the mosquito may have the same relationship to the mammal as the aphid has to the flowering plant?**

b   **What is the major difference between the aphid/flowering plant and the mosquito/mammal relationships?**

c   **What is the relationship between the mosquito and the malarial parasite?**

C   Gall wasp (or fly) and flowering plant
Many types of insect lay their eggs on parts of plants. When the young start to feed the plant may form growths which are characteristic of the species feeding on the plant, so that the size, shape, and distribution of the growths are an exact guide to the insect feeder. These growths are called galls.
1   Examine a gall, open it up, and observe the organism inside.

Questions

a   **What is the relationship of the insect to the gall?**

b   **How has the plant responded to the activities of the gall-former?**

c *Does the plant appear to have been harmed?*

D   Honey bee and flowering plant

1   Examine the legs of a worker honey bee (*Apis mellifera*) which have been mounted on a slide. Under the low power of the microscope try to identify the main features of each of the three pairs of legs.
2   Note the major differences in structure between the types of legs, and try to relate these differences to their functions.
3   Examine the structure of a typical insect-pollinated flower such as the white deadnettle (*Lamium album*). Look particularly at the size and position of the anthers and stigmas and their relationship to one another in space and in time.
4   Try to watch bees visiting such flowers and to work out the functional importance of the positions of anther and stigma.
5   Examine both the pollen grains and the stigma surface under the microscope.

Questions

a   *How would you describe the relationship of the bee to the flower?*

b   *From your examination of the pollen and the stigma, do you think that the plant is specialized for insect pollination in the structure of these two parts?*

E   Freshwater shrimp and Vorticella

Almost any freshwater crustacean will yield *Vorticella* attached to its exoskeleton. The bell-shaped head and the coiled stalk which contracts and expands are very typical. Specimens are also found on the surface of plants, and swimming in the water.

1   Observe the feeding methods of the crustacean and the *Vorticella.*

Questions

a   *What similarities are there in the feeding mechanisms of the two organisms?*

b   *What are the consequences of the type of behaviour observed?*

F   Lichens

1   Examine a sample of a lichen attached to a piece of wood. Note its surface features.
2   Examine a transverse section of the thallus and recognize the fungal and algal components.

Questions

a *What are the consequences of the arrangement of the fungal and algal components?*

b *How does the association between the two mutualists affect the types of habitat in which the lichens grow?*

INVESTIGATION
27F A grassland survey using sampling techniques

(*Study guide* 27.7 'Dispersal and other historical factors'; Study item 27.71 'Dispersal onto Surtsey'.)

A study of lawns, roadside verges, or pastures can illustrate some of the ways in which plants interact with their environments. Although the complete study of a community requires the identification of all the species, it is possible to simplify the task by concentrating on a small number of easily identified species.

In this investigation two adjacent areas of grassland are selected which differ in one major environmental factor that may affect the pattern of the vegetation. For example, grazing or cutting regimes may differ, they may be trampled to different extents, or the application of fertilizers may cause differences in the growth of some species. Random sampling of small areas within the areas to be compared is the most efficient method of comparison.

Procedure

1 Select an area of grassland where an environmental factor seems to be influencing the vegetation. A lawn that is partly shaded or trampled, or partly mown, would be suitable.

2 Place two long tapes at right angles along two of the margins of the sampled area.

3 Select the sampling points by using a table of random numbers. Choose a row of numbers, either horizontally or vertically, in the table. Use the first number to identify one co-ordinate (for example, E–W) and the other number to identify the other co-ordinate (for example S–N). Always place the quadrat frame in the same position relative to the grid.

4 Examine the sample area within the 50 × 50 cm frame. Identify the commoner species where possible. Unidentified species can be given a letter or description. Record your results in the form of a table, with the species listed down the left-hand margin and your estimates of abundance in columns to the right, one column for each quadrat.

---

5   The abundance can be estimated most efficiently, in terms of precision per unit time, by judging by eye the percentage cover of each species in each quadrat (see *Study guide II*, section 28.1). The percentage cover is the percentage of the quadrat area which is covered by a perpendicular projection on to it of the aerial parts of the species. It is estimated more easily if the quadrat has been gridded first with copper wire or string.

6   Record for each area the environmental factors which seem to influence the community such as light intensity, height of the grass, soil factors, and any other special features.

7   Test the significance of any differences between the results for the two areas of grassland by using a Mann–Whitney U-test, on the data for each species separately (see *Mathematics for biologists*).

8   Concentrate on the differences which are statistically significant. Explain how various physical or biological factors, such as mowing or trampling, might influence the distribution patterns of the species in the two areas.

Questions

a   *Describe the difference in the distribution pattern and numbers of plants found in the two areas.*

b   *Relate the difference in the two communities to differences in the environment. Suggest as many factors as you can which might account for the distribution pattern of the plants. Which may be the most important? Give reasons for your answer.*

c   *Which species seem to be*
   1   most tolerant and
   2   least tolerant
   of the environmental factors which you have listed above?

d   *Which features of these tolerant species might be responsible for their ability to survive?*

e   *Some of the plant species may exist in similar abundance in each habitat, but show considerable variation in growth form. Such species are said to exhibit plasticity if the variation is environmentally induced rather than the result of genetic differences between members of the same species. An example is shown in* **figure 12**. *Describe or draw any examples of plasticity which you observe in your areas.*

f   *Weeds can establish themselves after grass has been sown. How would seed size, seed numbers, speed and rate of germination, and*

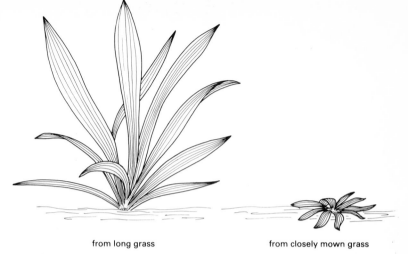

from long grass   from closely mown grass

Figure 12
An example of phenotypic plasticity in ribwort plantain, *Plantago lanceolata*. The phenotypes of these two plants are likely to differ because of the different environments in which they have developed. If the close mowing had continued in the same place for a long time, however, a low-growing race of the plantain might have developed which differed in genotype from the race in the long grass.

vegetative reproduction affect the establishment of weeds in grassland?

Suggestions for further reading

COLLINS, M. *Urban ecology*. Cambridge University Press, 1984.
DARLINGTON, A. and LEADLEY-BROWN, A. *One approach to ecology*. Longman, 1975.
DIXON, A. F. G. Studies in Biology No. 44, *The biology of aphids*. Arnold, 1973.
DOWDESWELL, W. H. *Ecology: principles and practice*. Heinemann Educational, 1984.
LEWIS, T. and TAYLOR, L. R. *Introduction to experimental ecology*. Academic Press, 1967.
Revised Nuffield Advanced Science: Biology. *Mathematics for biologists*. Longman, 1987.

The following is a useful aid to the identification of organisms:

DARLINGTON, A. and HIRONS, M. J. D. *The pocket encyclopaedia of plant galls*. Blandford, 1968.

CHAPTER 28

POPULATION DYNAMICS

INVESTIGATION
28A   Sampling methods for small invertebrates

(*Study guide* 28.1 'How can population size be assessed?')

It is seldom possible to count every individual member of a species in a habitat. An estimate of the population has to be made by counting sample areas. Methods involving search and capture of small animals can be unrewarding: some animals escape, some are hidden, and others visit from inaccessible places or from a distance, and are missed.

To sample active, nocturnal, and cryptic animals, traps can be used. The analysis of such captures can give clues to the time during which the animals are active, their relative numbers, and their rhythms of behaviour over long and short periods of time.

The aim of this investigation is to compare the animals in two different habitats, and to demonstrate the various methods of catching them. Four types of trap are available.

Pitfall trap. Plastic cups or similar containers are sunk into the ground so that their edges are flush with the surface of the soil, and any cracks between the jar's rim and the soil are filled with sand. The ground is firmed so that animals tend to fall into the container. A piece of wood or a tile is supported over the pot to keep out rain and debris; it also reduces the chance of interference. Baits of various kinds can be used: meat, dung, vegetable scraps, and so on. The alternative is to put a 2 per cent solution of methanal (formaldehyde) in the beaker to a depth of 3 or 4 cm. The position of each jar is marked with a cane or stick.

Cover trap. A large piece of wood or a tile is supported so that it forms a large, damp, sheltered area above some watered ground. House bricks soaked in water and placed cavity side down are a suitable alternative.

Water trap. Yellow paper is put into the bottom of a yogurt or ice-cream container, and water containing a little detergent or 2 per cent methanal is added to a depth of about 5 cm. (Yellow bowls are particularly effective traps for flying insects.)

Sweep-netting. Sample equal numbers of sweeps in both areas.

Procedure

1   Select two different habitats, close to one another, for comparison.
2   In each habitat, set up the traps in an area where they are unlikely to be disturbed. There is no need to set up all the different types of trap, but several traps of each kind are required. The types of trap, the numbers of traps used, and their distribution in the area will depend

on the type of investigation you select and the area in which you are working. Normally, you will need equal numbers of each type of trap scattered at random in each area. This will enable you to investigate differences between adjacent habitats.

3 Set up three traps of each type and leave for three days. Each day, check each trap and collect and total the catch. The traps must not be left for long, in case large carnivorous invertebrates in the traps eat some of the other animals. The cover traps will probably need six days, or more if the weather is dry, and should be left undisturbed for two days at a time.

4 Collect the catch from each type of trap using a pooter or a sieve, or, in the case of cover traps, a large etherizer. Transfer each catch to a separate numbered Petri dish and return to the laboratory.

5 Classify each individual as far as you can using the keys or identification guides available. The identification is time-consuming and you may find it easier to concentrate on four or five easily named species, comparing their relative abundances in the two areas. Record the results in a table: date and time of capture, position and catch in each trap, weather conditions, and so on.

6 Separate the scavengers, omnivores, herbivores, and carnivores as far as possible. Weigh each of these categories separately to provide a quantitative estimate of the standing crop at each trophic level. From these data you may be able to draw a pyramid of biomass for each habitat (see *Study guide II*, Study item 29.12).

Questions

a *Compare the types and number of invertebrates caught by the different kinds of traps. Which categories of animal only appeared in one type of trap? Suggest reasons for the differences and describe briefly how you could test one of your explanations experimentally.*

b *Which organism appeared in the traps most frequently? What information have you obtained about*
   1 its spatial distribution (where it is found),
   2 its temporal distribution (when it is captured),
   3 its population size (total numbers in the habitat)?
   What further experiments could be performed to confirm your impressions?

c *What proportions of carnivores, omnivores, and herbivores were captured in the traps? Does this reflect their proportions in the community?*

d *What differences appeared between habitats? Concentrate on the species which were found in several traps. Compare the numbers of each species in the two habitats by a Mann–Whitney U-test (see* **Mathematics for biologists***). For those species which were significantly more abundant in one habitat than another, try to explain their distribution patterns. You may be able to suggest what they eat, or how they select their habitat.*

INVESTIGATION
28B   Capture–mark–recapture technique

(*Study guide* 28.1 'How can population size be assessed?'; Study item 28.11 'The capture–mark–recapture method'.)

This technique is used to estimate the population size of small, mobile animals, such as grasshoppers, snails, woodlice, ants, pond-skaters, and small mammals. In its simplest form, some of the animals in a population are captured, marked, and released. After they have had enough time to mingle at random with the rest of the population, another group of animals is collected from the same place in the same way. Some of these recaptured individuals will have been marked, and the rest will not have been marked.

Let the number of animals captured, marked, and released be C.
Let the number of animals recaptured be R.
Let the number of marked individuals amongst those recaptured be M.
Then our estimate of the population size is CR/M; this expression is known as the Lincoln Index.

In simple numerical terms, imagine that you captured, marked, and released one hundred individuals (C = 100). A day later, using exactly the same procedure, you captured another hundred (R = 100) of which ten had been marked (M = 10). One-tenth of the individuals you captured on this occasion were marked individuals. The hundred marked individuals which you released probably represented one-tenth of the total population size. The population therefore is 1000 individuals. Or, in algebraic terms, CR/M = 10 000/10 = 1000.

Procedure
Carry out first a simulation exercise to illustrate the capture–mark–re-capture method. The population is represented by 'poppet' beads of one colour. Marked individuals are represented by substituting forty beads of a different colour for an equal number of the original beads. Thus C = 40.

1   Mix all the beads in an opaque bag and remove one handful. This will give a sample (R) of between 35 and 45.

2   Count up the number of marked beads and the total number of beads in the sample. The number of marked beads is M.
3   Calculate the population size (N) from the equation $N = CR/M$.
4   Replace the sample into the original population.
5   Repeat stages **1** to **4** several times, recording the estimate of the population each time.
6   Calculate the mean value for the estimated population size, and check your answer by comparing your results with those of the actual population size.

The model shows several important features which must be borne in mind when sampling real animal populations.

1 The size and constitution of the population should not change from one sampling to another. The beads are returned after sampling to simulate that the population being sampled has no births or deaths, and that no immigration or emigration has taken place.

2 The population should be found within finite boundaries, and should all be capable of being sampled.

3 Mixing of the marked and unmarked specimens should be rapid and at random.

4 The species must be abundant and easy to mark.

5 The marking system must not harm the organism, nor make it more obvious to the collector, nor affect the survival rate. The marks must persist for the duration of the experiment.

6 The sample must represent at least 10 per cent of the population.

7 The method of capture must not tend to recapture marked individuals; the whole population must be sampled whenever each sample is taken.

If a field investigation is carried out, it is important to take these criteria into consideration. Unless the population being sampled conforms to the criteria, methods must be used to allow for the differences found.

Such investigations can provide evidence of the changes in populations over a long period of time, once the technique has been shown to give consistent results, even if the results are not very accurate.

Procedure

1   In the available time collect and mark with non-toxic paint as many of the population as possible (C) from the defined area you have selected. Make sure that the boundaries of the area are clear.
2   Place the marked animals at the centre of the colony and allow them to disperse. Any adversely affected animals which fail to disperse should not be included in the original count.
3   After twenty-four hours, or a suitable period for the slower-moving animals, collect a further sample of the same size.

4 Count up the sample (R) and the marked individuals (M) and calculate the population size (N). N = CR/M.

5 Re-mark the sample with a different colour, and release them. By following the procedure outlined above, the population size can be assessed once more.

It may be possible to continue the exercise to monitor the changes in the size of the population and the causes of such changes.

Questions

a *What was the mean size of the population? Explain how you have arrived at your answer and give an indication of the confidence with which you express your answer.*

b *What are the principal errors encountered in your investigation? How did they affect the result, and how could you take account of them in estimating the true size of the population?*

c *What do you consider to be the principal practical use of an investigation of this sort? Consult the available literature and find an example where such an investigation has had important implications for humans.*

INVESTIGATION
28C **Population growth in *Chlorella***

(*Study guide* 28.2 'Population growth'; Study item 28.21 'The growth of a population of yeast'. *Study guide* 28.3 'Population limitation'; Study items 28.31 'The population dynamics of the great tit (*Parus major*) in Marley Wood, Wytham Woods, near Oxford' and 28.32 'Density-dependent, density-independent, and inverse density-dependent factors'.)

A rapidly dividing unicellular organism which reproduces asexually will demonstrate the major phases that occur during the growth of a population. It is difficult to demonstrate the phases in sexually reproducing organisms because the numbers increase irregularly. When the increase in asexually reproducing organisms occur in the field, the rate of population growth depends on a large number of interacting factors. In such cases as the development of algal 'blooms' on lakes there is the added complication that several species may be involved.

Practical problems include counting the numbers of the organisms involved, and relating the increase to the breeding population. The use of a unicellular organism allows the experimenter to develop a model of the phase of population growth. In this case, the use of a single species of alga (*Chlorella vulgaris*) reduces the problems.

Before carrying out this investigation you need to familiarize yourself with the use of a haemocytometer for estimating cell density.

Using a haemocytometer
Figure 12A illustrates the use of a haemocytometer slide. The slide and cover-slip are constructed so as to hold a suspension of cells as a thin layer over a graticule etched on the slide. The slide is strongly made, with

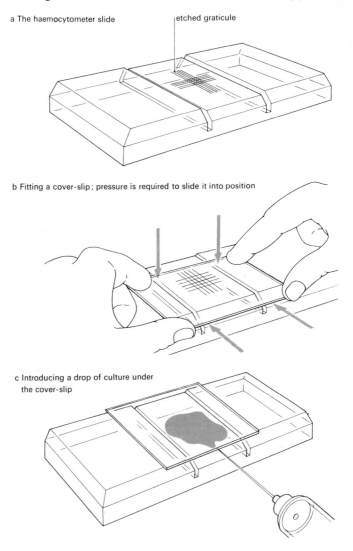

a The haemocytometer slide   etched graticule

b Fitting a cover-slip; pressure is required to slide it into position

c Introducing a drop of culture under the cover-slip

Figure 12A
Using a haemocytometer slide.

a central platform lower than the outer ones, and the slip is thick so that it resists bending. Thus we can make a film of cell suspension of known and uniform thickness lie over squares of known area. By counting the cells within a typical square, we can estimate their density. We can then compare the relative densities of different suspensions, and also calculate the absolute densities if we need to know these.

It is important to use the haemocytometer correctly. Practise the procedure as follows.

1   Examine the slide and see where the graticule is, with a hand lens. Read all the instructions before doing anything else.
2   Clean the slide and slip with a dry tissue or cloth.
3   Breathe on the slide to moisten it very slightly and then put the slip in position over the graticule by sliding it slowly onto the slide from the edge nearest to you. Keep it flat. Use both index fingers on the slip, pressing downwards firmly, but only where the slip is supported by the slide underneath. Move the slip across the slide by using both thumbs. Beware of breaking slips as they are expensive. One edge of the slip should project above the outer, vertical, edge of the slide.
4   Make sure that the slip is correctly attached to the slide by checking that rainbow-like patterns (Newton's rings) are visible over most of the area of contact. Repeat 3 if necessary. The junction between slide and slip should be sufficiently strong to keep the slide clinging to the slip when the slip is lifted.
5   Use a pipette or syringe to introduce the suspension as a drop below the edge of the cover-slip, filling the space between the platform and the slip almost completely. Do not over- or under-fill. There should not be any suspension in the moats around the platform.
6   Remove any surplus fluid from the top, sides, and bottom of the slide and slip.
7   Put the slide on the microscope stage and allow one minute for the cells to settle onto the graticule and become easy for you to count without focusing at different levels.
8   Light up the field of view under medium power by adjusting the light source, diaphragm, and condenser.
9   Find the graticule and adjust the magnification so that you can easily examine one large square without having to move the slide. Readjust the lighting if necessary.
10 Since the class will be comparing results each group must use the same standardized procedure for counting cells. Count the cells in four squares unless the suspension is very dense, when it may be sufficient to use four smaller oblong areas in a divided square. In this case remember to discover what proportion of a large square each oblong represents.

11 Some cells will lie on a boundary between squares. The usual procedure is to count the cells on the north and west boundaries and to ignore those on the south and east. It does not matter whether a cell is mostly in a neighbouring square; count it if it touches the north or west boundary. Otherwise ignore it.

12 Calculate the average number of cells per square so that you will eventually be able to plot the data in a graph.

Procedure

The size and cell density of the original population must be known if the experiment is to be replicated.

The method described here is suitable for a sample size of $10\,cm^3$.

1 Using a $1\,cm^3$ syringe, take a sample of $0.1\,cm^3$ from an actively growing dense culture of *Chlorella*.

2 Add the sample to $9.9\,cm^3$ water in the colorimeter tube.

3 Set the colorimeter to 100 per cent transmission using a blue filter ($410\,nm$) and a colorimeter tube containing water.

4 Replace the 'blank' with the sample and read the meter reading on the colorimeter.

5 Repeat stages **1** to **4** using $0.2\,cm^3$, $0.3\,cm^3$, $0.4\,cm^3$, etc., of *Chlorella* culture, making the volume up to $10\,cm^3$ each time. Record the colorimeter reading for each sample.

6 Determine the number of cells in a range of these samples by counting small samples using a haemocytometer. Several counts should be made at each cell density, to establish the level of precision of your procedure.

7 Plot the transmittance readings against the cell count for each of the dilutions. If the transmittance is directly proportional to the number of cells in the tube the graph will be a straight line.

8 From the graph, read off the highest cell count (horizontal axis) which gives 100 per cent transmittance (vertical axis). Find out what volume of *Chlorella* culture (see **5** above) needs to be made up to $10\,cm^3$ in order to provide this cell count. Multiply this volume of *Chlorella* culture by 5; this gives the volume V. (See *Practical guide 2*, investigation 5C.)

9 Remove volume V from the original *Chlorella* suspension and place it in a conical flask. Add a measured volume of culture medium to make the total volume up to $50\,cm^3$. Plug the neck(s) of the flask(s) with non-absorbent cottonwool. You are now ready to sample the growth of the population.

Sampling the growth of the population

10   Place the culture flasks about 70 cm from two 15 watt Gro-lux tubes
     and keep the flasks at about 20 °C for the period of the experiment.
11   Take samples of 10 cm^3 each day at about the same time. Determine
     the transmittance and return them to the culture afterwards. Use the
     graph of transmittance against cell count to find out the cell count in
     the culture on each occasion.
12   Continue the sampling and population determinations for two weeks.
     If there is any difficulty in sampling at the weekends it will be
     necessary to set up several cultures, each with the same initial density,
     on successive days, and to take samples so that a continuous and
     overlapping series of population estimates can be prepared. This will
     allow the mean of the population growth at all stages in its
     development to be calculated.

Recording the results

13   Use the graph of colorimeter reading against density of *Chlorella* cells
     to plot the number of cells against time, and then the logarithm of the
     number of cells against time. The use of log-linear graph paper allows the
     plotting of the results to be done directly.

Questions

a   *Does the evidence suggest that the growth of the population was
    logarithmic, arithmetic, or of some other kind? Name the period
    during which the growth was most nearly logarithmic.*

b   *What factors appear to be controlling growth in the conditions of
    your experiment?*

c   *In what way does the supply of energy to the* Chlorella *population
    differ from the supply of energy to most populations in the wild?*

d   *What factors could be altered to examine their effects on the pattern
    of growth of this population?*

e   *In what ways was it useful to use a colorimeter? What factors may
    cause inaccuracies in the method?*

INVESTIGATION

28D Population regulation in the water flea, *Daphnia*

(*Study guide* 28.3 'Population limitation'; Study item 28.32 'Density-dependent, density-independent, and inverse density-dependent factors'.)

Populations of organisms do not increase in numbers as fast as is theoretically possible, unless they are invading a new habitat which is particularly suitable. Most populations, at least in the absence of extreme human influences, fluctuate in size about a mean point.

The mechanisms which control populations are of two main types. First, there are those, such as climate or burning, which exert their random effects at all population densities. These are called *density-independent* factors. The second type have effects which become proportionately greater at high population densities. These are known as *density-dependent* factors. Density-dependent factors, for example, allow the use of insect parasitoids, in 'biological control', to regulate the populations of organisms harmful to humans (*Study guide II*, Study items 28.31, 28.32, and table 68).

These mechanisms are important for the maintenance of populations at optimum levels for feeding, reproduction, and growth. In this experiment the effects of density-independent and density-dependent factors on populations of the water flea, *Daphnia*, are investigated.

Procedure

In this experiment there are three treatments, known as A, B, and C. In treatment A, the control, the *Daphnia* population is allowed to grow unchecked. In treatment B, a random number of *Daphnia* is removed each week, whereas in treatment C, more *Daphnia* are removed the higher the size of the population.

All the experiments are in McCartney bottles to which *Daphnia*, yeast, and artificial pond water have been added. Each week, some *Daphnia* individuals are removed from treatments B and C to simulate death caused by density-independent and density-dependent factors.

The three treatments are as follows:

Treatment A. This is an unchecked population. After carrying out the weekly maintenance procedure (see below), simply return all the *Daphnia* carefully to next week's culture tube.

Treatment B. Density-independent control is brought about by the removal of a random number of *Daphnia* at each weekly maintenance period. After counting the total number of *Daphnia* in the population, select a random number from random number tables by starting on any vertical or horizontal line and finding the first random number lower than the number of *Daphnia*. Remove this number of *Daphnia* from the

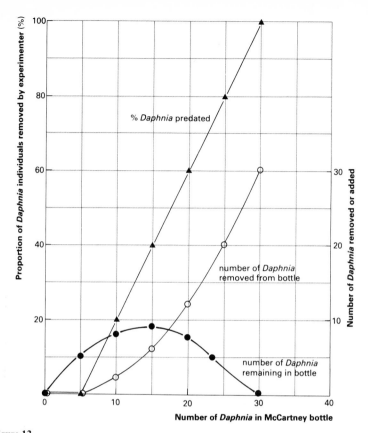

Figure 13
Density-dependent regulation of *Daphnia* in treatment C of the experiment. Each week, count the number of *Daphnia* individuals in each McCartney bottle. Knowing this number (horizontal axis) you can read off the vertical axis on the *right*-hand side of the graph   *i* the number of individuals to be transferred to the same bottle for the following week (those 'remaining in bottle') and   *ii* the number of individuals to be removed from the experiment and to be transferred to the stock culture instead (those 'removed from bottle'). This represents a percentage predation (straight-line graph, —▲—▲—) which can be read off the vertical axis on the left. (When five *Daphnia* individuals, or fewer, remain in a culture, no individuals should be removed.)

Petri dish and return them to the stock tank. Return the remainder to next week's culture tube.

Treatment C. Density-dependent control is brought about by removing a percentage of the population by reference to *figure 13*. Return the animals remaining after 'predation' to next week's culture tube.

1   Using a spirit marker, label fifteen McCartney bottles A1, A2, A3, A4, A5; B1, B2, B3, B4, B5; C1, C2, C3, C4, C5. Add your initials.
2   Add to each tube 12 cm³ of the artificial pond water provided.

3 Add to each tube 0.1 cm^3 of the yeast suspension, and mix well.

4 Add ten *Daphnia* to each tube, and screw the tops on.

5 Leave the tubes for one week. Examine them each day to check that the *Daphnia* are alive. If the yeast has settled, stir it up.

6 After one week, decant the *Daphnia* and the rest of the contents of each McCartney bottle into a separate labelled Petri dish. Count and record the numbers of *Daphnia* in each Petri dish.

7 Clean and rinse out all the McCartney bottles. To each, add 6 cm^3 artificial pond water from the stock bottle.

8 For treatments B and C calculate, for each of the ten McCartney bottles separately, the number of *Daphnia* which should be returned to each McCartney bottle (see above for a description of each treatment).

9 Using a wide-mouthed pipette, transfer the correct number of *Daphnia* from each Petri dish to its equivalent McCartney bottle. Do not transfer any debris from the old culture to the new one.

10 Add 0.1 cm^3 of yeast suspension to each tube, and mix well.

11 Add enough artificial pond water to make the volume of solution in each McCartney bottle up to 12 cm^3. Screw the top on to each bottle.

12 Leave the experiment for another week before recording the results again. Every week, carry out the maintenance procedures described in paragraphs **6** to **11** above.

13 Continue the experiment for about four weeks, or until the population in the McCartney bottle A begins to level off.

14 Plot the numbers of *Daphnia* against time for all fifteen bottles. To do this, create three graphs, with the same axes, one above another. On each graph plot the five results for each treatment.

Questions

a *Which treatment regulated the* **Daphnia** *population to a mean value?*

b *Which treatment was the least successful in regulating the* **Daphnia** *population to a mean value?*

c *Describe the kind of fluctuations in size which might occur in a population subject to density-dependent control.*

d *What does the variation in population density have in common with homeostatic mechanisms?*

e *What biological factors have been ignored or avoided in the experimental procedures suggested?*

f *What procedure could you use to eliminate or to make allowances for these biological factors in your experimental design?*

INVESTIGATION

The responses of a predator to changes in the number of its prey

(*Study guide* 28.3 'Population limitation'; Study item 28.32 'Density-dependent, density-independent, and inverse density-dependent factors'.)

Populations of organisms increase in numbers as reproduction takes place. Various factors control the size of the population but to be effective they must remove an increasing proportion of the population as the numbers increase. This kind of regulation is said to be density-dependent, and seems to operate in natural systems since population numbers usually fluctuate about a mean value.

One of the factors that controls the size of populations is predation, and this simulation exercise investigates the effects on a population as a single predator seeks a variable number of prey.

Procedure

1   Work in pairs, using one partner as the predator and the other as the recorder. Each 'predator' must sample all the 'prey' densities.

2   Collect a sheet of paper 60 × 60 cm and ruled in 2 cm squares. You will also need 225 discs. Anchor the paper firmly to the table top and place a collecting box next to the paper so that all discs collected can be placed in it without difficulty.

3   The 'predator' is blindfolded away from the sight of the board and the setting-up procedure. Meanwhile, the recorder places the discs in position on the paper. The number of discs used is selected at random from table 4, which also gives the average distance between adjacent discs. Place the discs in position at random. The whole experiment could be repeated later with the discs   *1* regularly arranged or *2* in clusters.

4   The predator searches for the discs for two minutes by tapping randomly on the sheet of paper with *one finger*. When a disc is found it

Number of discs on 60 × 60 cm grid	Density of discs on grid (numbers m^{-2})	Approximate average distance between centres of discs (cm)
9	25	20
16	44	15
25	70	10
50	139	7
100	278	5.5
225	625	4.0

Table 4
Numbers of discs to be used in the 'disc experiment'.

is picked up and placed in the box. After each capture, allow a time lapse by counting slowly up to three before continuing the search.

5   The two-minute search is repeated for each density, using the same predator.

6   Plot the numbers picked up (y) against the density (x), and collect the data from the whole class so that the mean values can also be plotted. Calculate the 95 per cent confidence limits for each of the results (see *Mathematics for biologists*). Plot them on the graph as vertical lines bisecting each point.

7   Calculate the percentage of prey 'captured' by the predator at each density. Plot a graph of these values against population density for your own results and the class results as before.

8   This investigation can be followed up by a real predator/prey experiment, such as predation by dragonfly nymphs on *Daphnia* at different densities, and the results compared with the results of the simulation.

Questions

a   *What happens to the number of prey killed as the population increases in size? Explain the shape of the graph at higher population densities.*

b   *In what ways does the graph showing the percentage of prey killed against density differ from the previous graph? What additional information does it provide?*

c   *What would happen to a population which was increasing in numbers if this model was correct?*

d   *Would this model fulfil the conditions given in the introduction, and give a density-dependent regulation?*

e   *Describe ways in which you could alter the model to achieve a better density-dependent regulation. Explain how the predators would achieve regulation at all densities.*

f   *What factors, other than predation, can exert density-dependent control on a population?*

Suggestions for further reading

Revised Nuffield Advanced Science: Biology, *Mathematics for biologists*. Longman, 1987.
WRATTEN, S. D. and FRY, G. L. A. *Field and laboratory exercises in ecology*. Arnold, 1980.

COMMUNITIES AND ECOSYSTEMS

INVESTIGATION
29A A quantitative study of an ecosystem

(*Study guide*. Study item 29.11 'Trophic levels and food webs'; Study item 29.12 'Ecological pyramids'.)

An ecosystem consists of the organisms living in a particular area, and the environment in which they live. The environment cannot be separated from the organisms, because it is constantly affecting them and in turn the organisms are affecting it.

Before an ecologist can study energy flow or nutrient cycling within an ecosystem, he or she needs to assign each organism in it to a particular trophic level. This is difficult for most natural systems because of the numbers of organisms in food webs, the complexity of their feeding interactions, and the problem of finding out what each species eats. As an illustration, look at the food web serving the herring (*Clupea harengus*) (*figure 14*). This is only one of the two hundred or so fish species to be found in the sea, and many large organisms exist there besides fish (can you think of some?). What a species eats depends on the time of day, the time of year, its size, and where it is. Furthermore, it is not easy to assign most heterotrophs to a single trophic level.

Here we shall use model aquatic systems because they are simpler and easier to study in the laboratory. The objects of this investigation are *1* to construct a food web for the community, *2* to assign each species to a trophic level, and *3* to construct a pyramid of biomass for the ecosystem. For identification, use keys to pond organisms. Remember, when you compare the findings of groups of students in a class, that ecosystems in different aquaria may not be exactly the same.

Autotrophs

The chart (*figure 15*) shows some of the autotrophs which are likely to exist in your system. You must identify and weigh both large and small autotrophs.

Procedure

1 Suspend a microscope slide in the water for at least three days to obtain a sample of autotrophic protists (*figure 16*). Clean its underside. Apply a coverslip and place it on the microscope stage. Obtain another sample from the side with a brush and pipette.

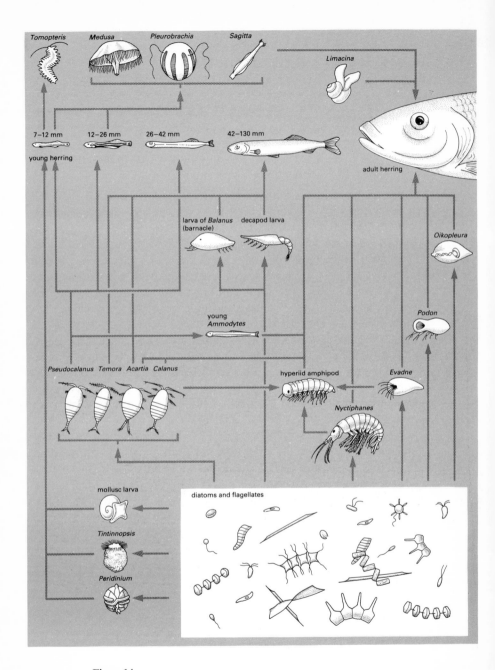

Figure 14
Food web of the herring (*Clupea harengus*).
Adapted from Hardy, *A.*, The open sea: its natural history. Part II. Fish and fisheries,
Collins, 1959.

	Species or other taxonomic grouping	Estimate of biomass in whole system
Higher plants		
Filamentous Algae		
Microscopic plants attached to sides		
Planktonic Algae		

Figure 15
The autotrophs in your model freshwater ecosystem. Tabulation of results.

Figure 16
Slides suspended vertically in a model ecosystem. The two further ones are held with paper clips over wire, the nearer one with cotton.
Photograph, Peter Fry.

2 Centrifuge a water sample to collect the plankton. Extract a drop of concentrate and transfer it to a cavity slide or to an ordinary one,

using a raised coverslip to avoid crushing the planktonic algae, single-celled organisms, and animals.

3   Identify each of the larger plants, count the number of individuals of each species, and measure the fresh biomass of each species. If you cannot immediately identify a species, give it a code name or make a sketch of it. Work methodically, disturbing the tank as little as possible, and make brief notes about any animals on the weeds, as a preliminary to your investigation of heterotrophs.

4   Make only approximate measurements of masses and numbers. A more precise approach will take too long. Express the results either as wet mass or the proportions of total mass collected.

5   Enter your results in a table, along the lines of *figure 15*.

Heterotrophs

Next we shall examine the organisms which may be feeding upon the autotrophs, the primary consumers, and the organisms which in turn may be feeding on them, the secondary consumers. Remember when you are considering the findings of several groups within a class that the species present in different aquaria are unlikely to be the same.

Procedure A. Herbivores

1   Watch the larger organisms in the aquarium to see which seem to be eating plants. A useful trick for observing beneath water is to use a beaker or Petri dish floating the correct way up and partly submerged (*figure 17*). Direct a strong light through the wall of the aquarium and observe the species from above.

2   Dissect out the foregut of a snail which you think has been eating plants. Obtain a sample of the contents of the gut, either by compression, using a glass rod as a rolling pin, or by rinsing with a gentle jet of water from a syringe. Examine the contents of the gut under a microscope and try to identify the material. Stains for cellulose and starch may help.

3   Alternatively, place a snail in a closed plastic container for twenty-four hours and examine under the microscope the faeces that accumulate.

4   You can remove the radula of a snail with forceps. Mount this in water under a coverslip to see how its structure limits the range of foodstuffs which a snail may eat.

5   If you have small mussels present you may suspect that they have been filter feeding by taking small particles in suspension. Isolate one in a dish of pond water, and allow the siphon to emerge. Place a drop of milk nearby to show water currents.

6   You may suspect that small animals such as *Daphnia* have been

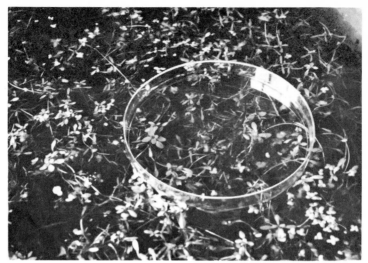

Figure 17
A Petri dish floating on the surface of a tank. This enables observers to watch activity beneath the water without their view being spoilt by reflection.
Photograph, Peter Fry.

feeding on planktonic algae. Examine the gut of one under a binocular microscope with good illumination.

7   Rotifers and ciliate protozoans are often found attached to leaves and floating debris. To watch these animals feed, supply them with a thick suspension of plankton. To prepare this plankton sample, centrifuge a sample of tank water, pour off most of the supernatant, and shake up the sediment with the remaining water.

8   Add your observations to your pond food web.

Procedure B. Carnivores
When the aquaria were set up some carnivores were purposely excluded. Some of them are so rapacious that they would kill all the herbivores in the aquarium within a few days. Amongst the most active predators are the larvae of dragonflies and beetles, and adults of beetles, bugs, and leeches. Some flying species may also have invaded the aquaria since they were set up. Make a note of these species.

1   Place a small piece of flesh in a specimen tube on the mud at the bottom of the aquarium at the start of the lesson. After half an hour, remove it to a dish of pond water and see whether there are planarians on it. If not, return the tube to the aquarium overnight. You can sometimes see the sucking proboscis of the planarian in action if you watch the feeding animal from below, through the bottom of a Petri dish.

2   Transfer a known number of herbivores of a single species to a specimen tube containing pond water. Add a single predator. Watch the predator for some time and record whether or not it tries to catch the potential prey. Leave the specimen tube overnight and record the next day how many herbivores have been eaten. This study can be extended, either by trying all possible combinations of predators and prey, or by placing a predator in a dish with a mixture of potential prey and seeing which herbivore(s) it catches.

3   The coelenterate *Hydra* feeds readily, particularly if it has been starved for a few days and can be viewed without disturbance.

4   Watch how arthropod carnivores use their mouthparts whilst feeding.

5   Incorporate these results in your food web.

Procedure C. Detritivores and decomposers

1   Watch planarians and *Asellus*, the water louse, to see if they appear to eat detritus. They probably feed on both living and dead material. Show this on your food web.

2   The main decomposers are bacteria and fungi. They feed mainly on detritus and on suspended organic material. Some mutualistic bacteria in the guts of herbivores and detritivores may also play an important role in decomposition.

Questions

a   *Does each species of herbivore feed on a specific range of plant species? Are different herbivores competing for the same plant food? If so, can you devise an experiment to test for competition? How do you think that each herbivore recognizes its food plants? How could you test experimentally your ideas about food recognition?*

b   *Do the carnivores have specific food preferences? Are there species which carnivorous predators do not catch and eat? Suggest some reasons why some herbivores are avoided as food.*

c   *Do the herbivores tend to feed for a greater proportion of the time than the carnivores? Can you explain your observations in terms of the problems in feeding faced by herbivores and carnivores in general?*

d   *Comment on the shape of your pyramid of biomass.*

e   *Is there evidence that the carnivores might control the population size of herbivores? Would the relationship between carnivores and herbivores be any different under natural conditions?*

INVESTIGATION
29B   The energetics of the stick insect (*Carausius morosus*)

(*Study guide* 29.1 'Introduction'; Study item 29.12 'Ecological pyramids'; and *Study guide* 29.3 'Energy flow'; Study item 29.32 Energy flow and agriculture'.)

The energy budget of an organism can be investigated by measuring its food intake and the ways in which the energy in the food is released and used. Carnivores may absorb and use 80 per cent of their food intake, but herbivores only assimilate about 40 per cent of the food they eat because of the high proportion of cellulose in their diets (see *Study guide II*, Study item 29.32).

In this investigation, the energy in the food eaten by a stick insect is estimated, and the fate of the energy determined. The energy budget of the stick insect during the week is represented by the equation:

$$FE = P + E + R$$

where FE = energy in food eaten
$\quad\quad\quad$ P = energy in new body mass and eggs (production)
$\quad\quad\quad$ E = energy in egesta
$\quad\quad\quad$ R = energy 'lost' in cellular respiration and other metabolic
$\quad\quad\quad\quad$ processes.

The energy content of exoskeleton cast in moulting is small and can be ignored.

Procedure
1   Select a healthy shoot of privet (*Ligustrum ovalifolium*) with about ten leaves. Trace the outlines of the leaves on graph paper without detaching them. Record carefully on the paper the positions of the leaves on the shoot. Weigh the shoot and record its mass.
2   Place the shoot in the apparatus (*figure 18*) and make sure that everything is prepared for the stick insect. A constant temperature cage should be used, if available.
3   Find the mass of a stick insect, preferably to three places of decimals, and place the insect inside the apparatus. Leave it there for a week.
4   After a week, dismantle the apparatus.
5   Remove the privet leaf blades one at a time, place them on the original tracing, and draw round them again. Calculate the area of leaf eaten and add up all the areas. Weigh the leaves and calculate the mass mm^{-2}. Calculate from this the mass of leaf consumed by the stick insect (M_1). Weigh the shoot and record its mass (this provides a check on the amount of food the insect has eaten, especially as some insects eat parts of the stem as well as the leaves).

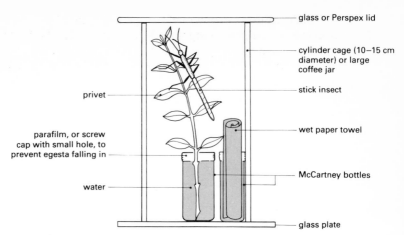

Figure 18
Apparatus suitable for the investigation of the energetics of a stick insect.

6   Reweigh the stick insect, and calculate its gain in mass (M_2).

7   Weigh separately two cavity microscope slides and label them 'eggs' and 'egesta'. Put the eggs on one slide and the egesta on the other and weigh them again. This is to provide some data if something subsequently goes wrong. Place these two slides, one with eggs and the other with egesta, in an oven at 105 °C for forty-eight hours. Reweigh them. Calculate the dry mass of eggs (M_3) and the dry mass of egesta (M_4).

8   Place the insect in a respirometer (*Practical guide 2*, page 18) and measure its respiration rate in terms of the volume of oxygen absorbed per unit time at room temperature. Carry out such determinations several times over a period of about one hour.

Calculations

1   Calculate the energy content of the food eaten (FE in the energy budget equation) by multiplying the mass of leaf eaten (M_1) by the energy content of fresh privet leaves ($4\,\text{kJ}\,\text{g}^{-1}$).

2   Calculate the energy content of the increase in mass of the insect by multiplying the increase in fresh mass of the insect (M_2) by the energy content of the fresh insect ($10\,\text{kJ}\,\text{g}^{-1}$).

3   Calculate the energy content of the eggs, by multiplying the dry mass of eggs (M_3) by the energy content of dry eggs ($24\,\text{kJ}\,\text{g}^{-1}$).

4   Calculate the energy content of the living material produced by the stick insect in a week by adding together the energy content of the increase in mass of the insect (calculation **2** above) and the energy content of the eggs produced (calculation **3** above). The result is P in the energy budget equation above.

5   Calculate the energy content of the egesta produced (E in the energy budget equation) by multiplying the *dry* mass of egesta (M_4) by the energy content of dry egesta ($24\,\text{kJ}\,\text{g}^{-1}$).

6   Calculate the energy lost in respiration in a *week* (R in the energy budget equation), assuming that as a consequence of the uptake of $1\,\text{cm}^3$ of oxygen, $0.02\,\text{kJ}$ of energy are released.

7   Now construct your energy budget equation, in kilojoules, as:

$$FE = P + E + R$$

Questions

a   *How closely do your results fit the equation? What are the main sources of inaccuracy?*

b   *Living organisms differ in the efficiency with which they convert their food material into biomass. The ratio of production (P) to respiration (R) provides a measure of food conversion efficiency. Calculate the P/R ratio for your stick insect and relate it to the values for other species given in table 5.*

Animal	Feeding method	P/R ratio
caterpillar	herbivore	1.4
grasshopper	herbivore	0.6
wolfspider	carnivore	0.5
perch	carnivore	0.4
elephant	herbivore	0.03
cow	herbivore	0.02

Table 5
Ratios of energy stored in production to energy lost in respiration in various animal species.

c   *What effect would a change in the mean temperature have on the conversion efficiency as estimated by the P/R ratio? Explain your answer.*

d   *How would you expect the energy budgets of young and old stick insects to differ?*

INVESTIGATION
29C   A study of decomposer organisms in the soil

(*Study guide* 29.3 'Energy flow'; Study item 29.31 'Decay and decomposition'.)

Some of the processes involved in decay are well known, but the organisms are difficult to observe. The use of artificial media enclosed in optically clear rectangular tubes (called micro-tubes) allows the direct examination of colonizing organisms with minimum disturbance. By using tubes of different sizes filled with different nutritive media, the pattern of decomposition can be studied.

Procedure

1 Fill several 10 cm pots with topsoil with a reasonably high humus content. Sieve the soil into a bucket or other suitable container, remove stones and debris, and mix it thoroughly.

2 Add sufficient water to make it moist but not saturated.

3 Place one micro-tube slowly into a beaker of molten agar and fill it. If necessary use capillator teats to suck up the agar. Remove the tube and place it across pieces of glass rod in a sterile plastic Petri dish. This should prevent most accidental contamination.

4 Repeat using a range of micro-tubes of different internal dimensions and a range of agars containing, for instance, soil extract, nutrient, malt, and cellulose. The tubes of size 0.3×1.2 mm, 0.6×2.4 mm, and 0.9×3.6 mm yield more rewarding results than smaller or larger ones do.

4 Secure the micro-tubes to microscope slides with very narrow strips of plastic insulating tape. Keep tubes of the same agar type together on the same slide and make a record of their contents.

5 Fill a 10 cm pot to about one-third of its capacity with the damp soil and place the slides vertically in the soil. Fill up the pots so that the soil covers the slides and place garden labels to indicate their position.

6 Water the soil with a fine spray to ensure that it is moist but not saturated, and weigh the pot and its contents.

7 Leave the pots for a minimum of two weeks in a warm place (about 20 °C) and add water with a fine spray to maintain the same mass throughout the whole period of the experiment.

8 Remove the tubes with great care and, after wiping the surface, examine them under the low, medium, and high power of the microscope, as appropriate. Use phase contrast if it is available. If observations are intended to last several days, plug the ends of the tubes with a little Vaseline to prevent drying out.

9 Record the types of organism found in each of the different tubes. It is not necessary to identify all the organisms, but each distinct type should be drawn accurately, its size shown, and a code letter of some kind assigned to it for future reference.

10 Tabulate your observations, recording the tube size, the type of medium and the organisms found. Tubes can be kept in a refrigerator overnight if counting and tabulating take too long.

11 Observe the organisms closely and try to find examples of movement, feeding, or other activities which may allow you to determine the types of relationship that exist in the natural environment.

Questions

a *What other environmental factors, apart from a high soil humus content, would create conditions suitable for these soil organisms?*

b *Devise an experiment, using this technique, to investigate the effect of one of these factors on the variety of organisms found in the microtubes.*

c *In what ways is the investigation likely to provide a restricted idea of the decomposition processes in the soil?*

d *Compare the types of organisms you have found in the different media. Which of the media contains*
  1 *the largest variety of organisms,*
  2 *only fungi and the herbivores that feed on them,*
  3 *mainly bacteria,*
  4 *soil amoebae, and*
  5 *eelworms?*

e *Explain how the size of the tubes affected the variety and numbers of organisms found in the tubes.*

f *Construct a food chain or a food web from your observations on the tubes. Look for interactions between the organisms involved.*

INVESTIGATION
29D A comparison of the growth of tolerant and non-tolerant seedlings when exposed to metal ions

(*Study guide* 29.5 'Pollution'.)

The heavy metals copper, zinc, lead, and nickel are present in certain soils as a result of natural outcrops of ore-bearing rocks or from the presence of mining or smelting waste. Most plant species cannot grow in soils containing heavy metals but studies begun in 1934 have shown that some species have become tolerant to the presence of these polluting ions in the soil. The commercial development of tolerant varieties can be of great importance in reclaiming such land and preventing soil erosion.

Procedure

1 Fill ten beakers (or shallow plastic ice-cream boxes or margarine pots) with normal culture solution. Cover the solution in the containers with

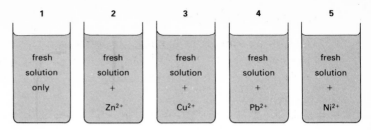

Figure 19
Culture solutions for tolerant and non-tolerant seedlings.

alkathene beads or perlite, or with nylon stocking which just touches
the solution.

2   Sow 'seeds' of the two varieties of creeping red fescue (*Festuca rubra*)
called Dawson and Merlin on the perlite or nylon stocking. Five
containers for each variety are required, with about ten 'seeds' evenly
spaced in a row over each one. Each beaker must be clearly labelled
with the name of the treatment. In case up to half the seeds fail to
germinate, germinate some spare seeds elsewhere.

3   To reduce evaporation you may have to keep all the containers moist,
for example beneath an upturned aquarium.

4   Allow 7 to 10 days for germination (much longer in winter). Then,
without removing it from the perlite or stocking, measure the length of
the longest root on each seedling. Make sure that you record the root
lengths from one end of the row of seedlings to the other, and that you
note at which end you began to record. Later on, you will have to
record again the root length of *each seedling*.

5   Replace the seedlings, to allow further growth.

6   After four more days remeasure the longest root on each seedling.

7   Transfer all the seedlings to new culture solutions, as shown in
figure 19.

8   After a further four days measure each root as before.

9   Subtract the original length of each root from the final length to assess
how much each root has grown. Calculate the average increase in root
length for each variety in each solution.

10   Calculate the Index of Tolerance (I.T.) for each variety and for each
metal ion:

$$\text{I.T.} = \frac{\text{growth during 4 days in culture solution} + \text{metal ion}}{\text{growth during 4 days in culture solution} - \text{metal ion}} \times 1000$$

11   Carry out Mann–Whitney U-tests (see *Mathematics for biologists*) to
compare
    1 the growth of the two varieties in the same ionic solution
    2 the growth of the same variety in different solutions.